BIRD DROPPINGS

A FUNTASIA OF FEATHERED FARCE

by Pete Jacob AND Barry Knowles

Dalesman Books – 1988

The Dalesman Publishing Company Ltd.,
Clapham, Lancaster, LA2 8EB.
First published 1988

ISBN: 0 85206 933 2

Dedicated to
Pat and "Joe" Louis,
with love and thanks.

Printed by Fretwell & Cox Ltd.,
Goulbourne Street, Keighley, West Yorkshire BD21 1PZ.

BIRD DROPPINGS

BIRD DROPPINGS

"By Jove! It's Lord Knowles — we haven't met since silent film days . . ."

"Sir Peter! Join me on this park bench — I'm trying to dream up an idea for a new book . . ."

"It looks a remarkably mucky bench . . ."

"It's the starlings I fear . . ."

"So that's not Snowcem on your shoulder?"

"What? Oh drat!"

"Get your own back, my dear sir — do a book about birds . . ."

"Can you think of any ideas?"

"If we repair to a public house and you can procure me a beer mat and a pencil, I'll have a go . . ."

"On your own head be it . . ."

"If we don't get out of this park soon, it probably will be . . ."

Pete and Barry

BIRD DROPPINGS

BIRD DROPPINGS

BIRD DROPPINGS

"I wish you'd buy a new tin opener, mate . . ."

BIRD DROPPINGS

BIRD DROPPINGS

"Dammit! Here's another one gone over to cartons . . ."

BIRD DROPPINGS

"Good grief! I think that chap's a talent scout for a duvet company . . ."

BIRD DROPPINGS

"I've always wanted to splash out on a new Rolls . . .!"

"Can't stop — I've a to go and pose on a yule-log for a Christmas card company . . ."

BIRD DROPPINGS

"Good news — we could be in for a killing on Wall Street . . ."

"If that really is Bob Geldof on Nelson's column we might still be in with the chance of a snack . . ."

BIRD DROPPINGS

BIRD DROPPINGS

STUART HALL'S ADVICE TO THE HAIRLESS
BALD EAGLE

"I'm not voting for the Blue Tit party again — they want to flog off all the council nests . . ."

BIRD DROPPINGS

"Damned inflation . . . !"

"What do you think, lads — is it a scarecrow or Patrick Moore . . .?"

LOVE BIRDS
DIVORCED
Barry Knowles

Kelloggs
CORN CRAKES
SHREDDED TWEET
WEET-A-BEAKS
QUACKER OATS
BarryKnowles

"Here's a nice little chimney at a reasonable price. Unfortunately it has a sitting tenant . . ."

"Watch the chappie!"

BIRD DROPPINGS

BIRD DROPPINGS

"We can't afford to migrate to Africa this year but we could manage a weekend in Bognor . . ."

"Which end is the cream . . .?"

BIRD DROPPINGS

"It's a fine looking egg all right, but pride comes before a fall . . ."

BIRD DROPPINGS

"You are not a great auk — you're just a filthy old bustard . . ."

"What makes you think you've got a future in show business . . ."

BIRD DROPPINGS

BIRD DROPPINGS

"Blimey! With friends like Prince Philip, who needs enemas . . .?"

"I've just had a great rub-down with a sandpiper . . ."

"Aww — come on — one good tern deserves another . . ."

"Here comes that mincing old Queen Penguin . . .!"

BIRD DROPPINGS

"Don't move a muscle — Bernard Matthew's got his eye on us now . . ."

BIRD DROPPINGS

"Congratulations men! We've wrecked the flight plans of a Boeing 747, 2 Starfighters and David Frosts's clapped-out old Concorde before breakfast . . ."

BIRD DROPPINGS

"It's not snowing at all — those damn house martins have been flying over again . . ."

BIRD DROPPINGS

"Not on ME — you silly tweet . . .!"

THE
UNTOUCHABLES
WITH
ELLIOT NEST

"If you know of a better pole — go to it . . ."

BIRD DROPPINGS

BIRD DROPPINGS

"What do you think of my flying Peter Scotts?"

"That's what happens when you turn vegetarian . . ."

"How long have you had this feeling that you're Concorde . . .?"

"I wish you'd stop buying junk food . . ."

BIRD DROPPINGS

"Congratulations — it's an omelette . . .!"

"I don't actually think they are practising yoga . . ."

"Look, dear — they've named it after you . . ."

BIRD DROPPINGS

"Run for it, Hildegarde — it's that man from Paxo . . .!"

"Our family's great claim to fame is that your grandfather was the first to splatter Tippi Hedren . . ."

BIRD DROPPINGS

BIRD DROPPINGS

"Have a nice lay now . . ."

"Mark my words, Doris, today we are *__not__* *a protected species . . ."*

"Old George is a frustrated circus performer . . ."

BIRD DROPPINGS

"Clear off cuckoo — this is not a timeshare nest . . ."

BIRD DROPPINGS

"Quite right too — damned messy things . . ."

BIRD DROPPINGS

"Millet's off, cuttle-fish is off and we've run out of worms . . ."

BIRD DROPPINGS

"That's shown 'em they're not the only ones who can cause havoc with fallout . . .!"

BIRD DROPPINGS

"You great twit! I told you you couldn't hatch a Cadbury's Creme Egg . . .!"

BIRD DROPPINGS

"You've got to let me have my evil way with you, or we'll become extinct . . .!"

"I can't sit down today, I'm having terrible trouble with my parson's nose . . ."

TODAY:
CROSS
CHANNEL
WADE
Barry Knowles

"Let's migrate to Brixton . . ."

BIRD DROPPINGS

"Are you really nesting in David Bellamy's boot?"

"We'd be well off if they paid us royalties on this margarine stuff . . ."

"It's no wonder the old twit's got gout . . ."

"Oh no! That's all we damned well need . . ."

"I'm afraid the little fellow's got worms . . ."

BIRD DROPPINGS

BIRD DROPPINGS

"Stop laughing — it's no fun being egg-bound . . ."

BIRD DROPPINGS

"Avert your eyes, Evadne — it's that dirty old fowl pest again . . ."

"He's what they call the tern of the century . . ."

BIRD DROPPINGS

"Ignore him — he's a raven lunatic . . ."

BIRD DROPPINGS

"Shut up, blabber-beak — brevity is the soul of twit . . ."

BIRD DROPPINGS

"It's disgraceful at his age — going off for a dirty weekend with his secretary bird . . ."

"Look the first cuckold of Spring . . ."

"If I were you I'd go on a crash diet before he starts thinking about Christmas . . ."

"I knew it was a mistake going on that fertility drug . . ."

BIRD DROPPINGS

"Personally I think a birdbath jacuzzi is a damn silly idea . . ."

BIRD DROPPINGS

BIRD DROPPINGS

"I've always been a fan of Barnes Wallis . . ."

"It's no wonder you're overweight. Scavenge outside a fruit and veg stall for a month . . ."

"I see she's found a new plover boy . . ."

"Home-wrecker . . . !"

"Don't worry — it's not AIDS — just a mild dose of chicken pox . . ."

"Now you know why they call him a nightjar . . ."

"Tragic really, he survived acid rain, nuclear fallout, ozone layer destruction and low-flying aircraft and then choked to death on his breakfast worm . . ."

"We were lucky to get an egg-sitter for tonight . . ."

"Move that ladder, polish the bell, clean the mirror and sweep up all those seed husks . . ."

BIRD DROPPINGS

"Let's see how they like being stared at . . .!"

‘It’s alright for you — I couldn’t get a big enough mortgage . . .”

"Greedy devil! Why didn't you tell me she was dishing out the 'Swoop' . . .?"

BIRD DROPPINGS

"George! Look — we've got a double yolker . . ."

"Sorry I'm late — I got caught up in an unexpected thermal . . ."

BIRD DROPPINGS

"That should discourage him for a bit . . .!"

BIRD DROPPINGS

"He was the only bird to splatter Peter Scott, David Bellamy and Bill Oddie in a single day . . ."

"Oh my Gawd! Look what she's bought . . ."

BIRD DROPPINGS